BLAME IT ON
EL NIÑO

BLAME IT ON
EL NIÑO

By Susan Dudley Gold

Expert Review by
David Adamec, NASA oceanographer

RSVP

RAINTREE
STECK-VAUGHN
PUBLISHERS
A Steck-Vaughn Company

Austin, Texas
www.steck-vaughn.com

Dedication
To the members of the LSC&TS: Carol Rancourt, Sharon Prosser, Karen Leary, Earlene Kitty Chadbourne, and Suzan Nelson. Your friendship and support enrich my life.

Published by Raintree Steck–Vaughn Publishers, an imprint of Steck–Vaughn Company

Acknowledgments
With grateful appreciation and thanks to David Adamec, NASA oceanographer, who generously reviewed the manuscript and provided graphics and information, and whose advice and expertise were invaluable. He improved the book immensely.
Thanks, as well, to Jill Cournoyer; Cynthia A. Howell and Robert E. Gabrys at NASA's Goddard Space Flight Center; NOAA's Mark Eakin; the NOAA/PMEL/TAO Project Office, Linda Stratton, and H. Paul Freitag for their assistance in this project; and to Ann Gibbs of the USGS, who graciously spent time and effort in tracking down photographs for the book.

Library of Congress Cataloging-in-Publication Data

Gold, Susan Dudley.
    Blame It on El Niño / by Susan Dudley Gold; expert review by
David Adamec, NASA oceanographer.
      Includes bibliographical references and index.
      Summary: Examines the weather phenomenon known as El Niño, common misperceptions about it, and the results of years of scientific study about it.
      ISBN 0-7398-1376-5
      1. El Niño Current Juvenile literature. [1. El Niño Current.]
    I. Title.
GC296.8.E4G65. 1999
551.6 — dc21   99-21660

Printed in Mexico.
1 2 3 4 5 6 7 8 9 0 LB 04 03 02 00 99

Design, Typography, and Setup
Custom Communications, Saco, ME

Illustration and Photo Credits
Cover: front left © Corbis, *front right* and *center* © PhotoDisc, Inc. *Cover, back* © PhotoDisc, Inc.
Inside: pp. 3, 6, 9, 11, 13, 26, 33, 42, 48, 58, 63, 67, 70, and *75* © PhotoDisc, Inc.; *pp. 5, 24,* and *34* courtesy National Oceanic and Atmospheric Administration (NOAA); *pp. 19, 22 (left* and *right), 45, 57 (left, center,* and *right), and 65* from NASA's Goddard Space Flight Center poster; *pp. 20 (top* and *bottom), 51, 52,* and *54* courtesy NOAA/PMEL/TAO Project Office, Dr. Michael J. McPhaden, Director, *p. 51* photo by Linda Stratton; *pp. 28* and *31* © Susan Gold based on graphic from NOAA's Climate Diagnostics Center; *p. 37 (top* and *bottom),* courtesy of NOAA's Climate Prediction Center; *p. 40* © Susan Gold based on data from *The New York Times,* graphic courtesy National Aeronautics and Space Administration (NASA); *p. 47 (left),* courtesy NASA's Jet Propulsion Laboratory and *(right),* courtesy NASA's Aerospace Education Services Program; *p. 60,* GOES satellite, courtesy NASA; *p. 64* © Susan Gold based on graphic from NOAA's Office of Global Programs; *p. 69,* NOAA-14 Polar Orbiting Environmental Satellite, courtesy NOAA; *p. 73 (top* and *bottom),* courtesy U.S.Geological Survey.

CONTENTS

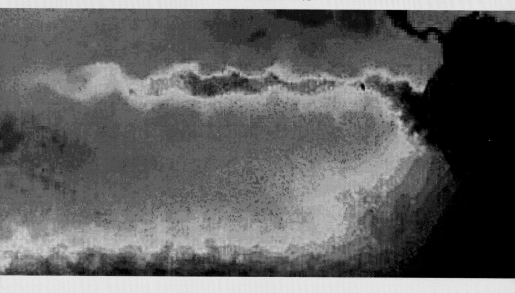

EL NIÑO: FACT AND FANTASY

*E*l Niño, the powerful ocean event that origi-
nates in the Pacific Ocean, has been blamed for
everything from lost socks to longer days. Sur-
prisingly, many of the claims are true—or at least they
have some link to El Niño.

An annoyed ABC News viewer wrote in to com-
plain that El Niño was at the root of everything. He
said—tongue in cheek—that he had lost his socks and
had failed to win the lottery because of El Niño.
Humorist Dave Barry claimed El Niño was to blame
for several errors he had made in recent columns,
including his mistaken placement of Harvard University in Princeton, New Jersey.[1] A *Boston Globe* columnist
accused the event of having "strange effects" on Mas-
sachusetts politics.[2]

Other people have blamed El Niño for giving them
the flu or causing their favorite sports team to lose the
championship game.

Actor Billy Crystal suggested that the ocean event
prevented Barbra Streisand from getting an Oscar.
Even Kathy, the daily comics page's most persistent
weight watcher, pinpointed El Niño as the source of
her woes. El Niño, she complained, had made all her
pants shrink.

Clearly, these events can't logically be blamed on El
Niño. But some of the real effects of El Niño are almost

as bizarre. Here are a few of the unusual occurrences that have been linked to El Niño:

- Torrential rains tied to the 1997–1998 El Niño transformed Peru's Sechura Desert into a lake 90 miles (144 km) long, 20 miles (32 km) wide, and 10 (16 km) feet deep, the second largest lake in the country.
- El Niño rains washed smog from the atmosphere over Los Angeles, resulting in the cleanest air in that region in 50 years.
- A wet, cool El Niño spring led to a population explosion of rats in New Mexico in 1983; fleas on the rats spread bubonic plague to people, and 24 died.
- People living in the northeastern United States enjoyed the warmest winter ever recorded and cut their fuel bills to the tune of $5 billion during the winter of 1997–1998.
- West Virginia residents, digging out from under 32 inches (81 cm) of snow in January 1998, braced themselves as another foot of snow fell on their communities. Townspeople lost power for days; the heavy snow collapsed the roofs of more than 100 buildings.

El Niño has brought unusual weather patterns to

El Niño, the powerful ocean event, brings both good and bad to regions around the world.

regions around the world for centuries. During El Niño years, nations along the western coasts of South America and North America endure torrential rains, floods, and mud slides. A typical El Niño brings mild winters to the northern region of the United States and Canada and droughts to southeastern Asia, southern Africa, northern Brazil, and Australia. El Niño's effect on global weather has been blamed for out-of-control wildfires in the Amazon rain forest, massive floods in China, more hurricanes in the Pacific Ocean, and fewer hurricanes in the Atlantic.

In some regions, El Niño has been a force for both good and bad. While anchovies that usually thrive in the cold waters off Peru migrate to Chile during El Niño years, sardines and shrimp are drawn to Peru's warm waters. Inhabitants of desert lands have traditionally welcomed El Niño's rains. Wolf Arntz, a German researcher, noted in 1894 that heavy rains filled Peru's coastal desert with a "hitherto unknown outburst of vegetation that covered wide areas with carpets of flowers for several months, enabling local settlers to raise cattle, sheep, and goats."[3]

one

WHAT IS EL NIÑO?

*O*n January 8, 1998, Jill Cournoyer awoke to a world of ice. El Niño's warm winter had suddenly turned harsh. Overnight, a catastrophic ice storm had brought the entire northeastern United States and southeastern Canada to its knees. Outside Cournoyer's Portland, Maine, home, ice coated everything: the tiny tendrils of bushes, pebbles in the driveway, road signs, and pavement. Her ice-encrusted car looked as if it had been stored in a freezer. Branches of all shapes and sizes littered the ground. A tree blocked the road. The trunk of another tree, split down the middle, lay on its side in her backyard.

"It looked like a tornado had torn through," said

Cournoyer. "My neighbor's tree fell over in front of me while I was standing in my driveway. A big branch bounced onto my shed roof. Ice was all over the ground."[1]

Nights were worse. Cournoyer remembers sitting in her cold, dark house not knowing when a falling tree would crash through her roof. "You could hear the trees cracking and the branches breaking off," she recalled. "It was so loud I couldn't sleep, a constant cracking and popping. Every time the wind blew, the trees creaked under the ice."[2] The monotonous hum of a generator here and there could be heard in the distance. Other than that and the sound of the trees, it was dead quiet.

Power lines throughout the region snapped under the weight of the heavy ice. In some spots, poles broke in two and dragged down the lines connected to them. A second storm wiped out power in coastal Maine.

Four-fifths of the state's residents—more than 965,000 people—lost electrical power, some for more than two weeks. Southern Quebec suffered even more devastation from the ice storm. By the fifth day of the storm, ten people had died in Quebec, and three million were without power for weeks. Many had no phones, no heat, and no water. Live wires blocked traffic. Sections of at least 50 highways in Maine and many more in Quebec had to be blocked off because of debris and fallen wires. Radio and TV stations went off the air as ice interfered with their transmission towers. Utility workers from as far away as North Carolina came to help repair the downed electrical systems.

Many people moved out of their homes into shelters, stayed with friends or relatives, or rented hotel rooms.

Ice storms in the northeastern United States caused millions of dollars of damage as a result of the 1997–1998 El Niño.

Others, however, made the best of things, operating generators or wood stoves for heat and cooking over gas grills and fireplaces. They used melted snow to flush toilets. In some areas schools closed for weeks. Families entertained themselves playing cards, reading by candlelight, and singing songs.

Like other El Niño-related events, the 1998 ice storm took its place as one of the region's worst natural disasters. In Maine, storm costs soared to more than $100 million to repair the state's electrical system, rebuild homes, run emergency shelters, and replace spoiled food. Four people died in storm-related accidents, and thousands were injured.[3]

Canadians were expected to seek $1 billion in insurance claims as a result of the storm in Quebec.[4] It was, according to Environment Canada, "by far the costliest weather catastrophe in Canadian history."[5]

Even though it is difficult to tie any one particular event to El Niño, most scientists agreed El Niño was to blame for these storms.

STRANGE WEATHER

Every two to seven years, a large pool of warm water builds up in the Pacific Ocean along the equator near the west coast of South America. The warm water heats the air above it, affecting the air pressure, the sea level, and the winds that blow across the Pacific. These, in turn, affect the weather, causing heavy rains in dry areas and droughts in places generally humid.

Local fishermen originally gave the name El Niño—Spanish for small boy or the Christ Child—to a warm current of water that appears off the west coast of Ecuador around Christmastime. The appearance of the warm water signaled the beginning of the fishing season, a blessing for the local people. During the 1920s, scientists adopted the name to describe the large-scale event that we associate today with El Niño.[6]

Researchers have studied coral, ice, soil, and trees to trace past El Niños—in some cases as far back as two million years. The event is mentioned in ships' logs and travelers' accounts as early as the 1500s. Historians speculate that El Niño may have aided Francisco Pizarro's conquest of Peru. When the Spanish conquistador landed in northern Peru in 1531, the usually barren lands yielded plenty of food and water for his troops and horses thanks to El Niño's rains. El Niño weather had its downside for Pizarro as well; the standing pools of water in Chira Valley, Peru, provided ideal conditions for the mosquitoes that spread malaria. The disease took a heavy toll on Pizarro's forces.

Because it lies right in the path of El Niño, Peru is one of the first nations to feel its effects. Heavy rains hit the normally arid coastal regions, flooding some areas and turning desert lands into lush gardens in other areas. A major El Niño in 1891 devastated Peru and captured the world's attention. Federico Alfonzo Pezet, a Peruvian geographer asked his colleagues at the 1895 International Geographical Congress being held in Lima to undertake "studies, surveys, and observations" to "get to the bottom of the question" of Peru's strange weather.[7] That same year, Alfred Sears, in a lecture to the American Geographical Society, described "septennial rains" that he said occurred every seven years in northern Peru. During that time, Sears told the audience, the desert was transformed into a garden where "lifeless earth springs into being; … flowering plants appear on every hand, grown to the height of a horse's head."[8]

During this time, a few scientists began to see links between unusual weather in one region and storms in other parts of the world. In the late 1880s, Charles Todd, a meteorologist in South Australia, noted that India and Australia suffered from droughts at the same time. He thought the atmosphere might carry the dry conditions from one location to the other. Henry Blanford, a weather reporter in India, suggested that heavy snows in the Himalayas might be linked to a lack of monsoons that brought rain to India.

Not until the twentieth century did scientists realize that El Niño might affect weather conditions in other regions of the world.

FINDING A PIECE OF THE PUZZLE

In the 1920s, British meteorologist Gilbert Walker was in India trying to find out why the area's monsoons—winds carrying rains to nourish the dry lands there—periodically failed to appear. When that happened, crops withered, and many people starved to death. While examining records of world weather patterns, he discovered a strong link between the air pressure in Darwin, Australia, and that in Tahiti. When the pressure off Australia was low, Walker noted, the pressure off Tahiti was high. The reverse was also true.

Walker called the pattern the Southern Oscillation (SO). When pressure is high in the east (near Tahiti), it causes strong winds to blow along the equator from east to west toward Australia and Indonesia. When the pressure pattern reverses—high in the west and low in the east—the prevailing easterly winds weaken and can even reverse, blowing toward South America.

Walker's studies showed that droughts often hit Australia, Indonesia, India, and parts of Africa during years when the winds blew from the west. Walker also noticed that during those same years, western Canada enjoyed especially mild winters. Could there be a link? Walker thought so. But he didn't have enough information—or the technology to collect it—to prove his theory.

Scientists of the time doubted Walker's views of a world weather system. They believed that weather was local; conditions in one area couldn't possibly affect other places across the globe. When Walker died in 1959, his obituary in the *Journal of the Royal Meteorological Society* gave a disparaging report of his beliefs:

> Walker's hope was presumably not only to unearth relations useful for forecasting, but to discover sufficient and sufficiently important relations to provide a productive starting point for a theory of world weather. It hardly seems to be working out like that.[9]

It took several decades, but Walker's groundbreaking research eventually provided a key piece to the El Niño puzzle.

SPOTLIGHT ON PERU

A major El Niño in 1957–58 once again focused the world's attention on Peru. Peruvian farmers had long collected the waste (guano) of coastal birds and sold it to buyers in Europe and North America for use as fertilizer. The "guano birds" played a big role in the nation's economy.

When El Niño of 1957–58 hit, fish that fed on the nutrients found in the cold water along Peru's coast went elsewhere as the water temperature rose and the flow of nutrients was cut off. Unable to find enough fish to feed on, the guano birds starved in great numbers. In a region where 30 million birds once produced guano, only 16 million birds survived.[10] The loss of the birds spelled disaster for Peru's economy and severely reduced fertilizer supplies in the countries where guano was sold.

As it happened, the nations of the world had agreed that year to join forces to study the earth, the atmosphere, and the oceans. The Scientific Committee for Oceanographic Research drew experts from many fields: meteorologists, scientists who study the weather; oceanographers; atmospheric scientists; biologists; hydrographers, experts who study and map bodies of water; and others. The appearance of El Niño and its devastating effect on Peru kindled the scientists' interest in the mysterious event. As a result, the committee collected much data on the 1957–1958 El Niño. That information provided Jacob Bjerknes, an atmospheric scientist at the University of California at Los Angeles, with the tools he needed to explain how El Niño worked. Bjerknes identified swings in sea levels similar to the changes Walker had recorded in air pressure. Using Walker's model, Bjerknes concluded that changes in winds, ocean temperature, and sea pressure in the Pacific region were all linked. When the sea and air pressure reversed, he noted, El Niño emerged, affecting weather conditions worldwide.

HOW EL NIÑO WORKS

Sun, rain, wind, and sea are inextricably linked as they circle the globe in a rhythmic dance. The sun begins by heating low-lying air in the tropics. The warm air rises, and as it does, the colder, heavier air over the North and South poles flows toward the equator. The rotation of the earth pushes the polar air away from the equator. In the Northern Hemisphere the air is forced to the right in a clockwise motion, while south of the equator the air twists counterclockwise to the left. The winds created by these events are called the trade winds.

As the trade winds travel along the tropical Pacific, they push the warm, top layer of the ocean away from the west coast of South America and toward the coasts of Australia and Indonesia. This mass of warm water, moved by the wind, raises the level of the ocean in the western Pacific. The ocean is nearly one-half meter higher off Indonesia and Australia than it is off Peru.

As the warm top layer is blown off the coastal regions of western South America, colder water from below comes to the surface. Rich nutrients and marine life thrive in

The chart at right shows the dramatic increase in water temperatures in the eastern Pacific as El Niño travels along the equator toward the western coast of South America. Red indicates warm water; blue is cold water.

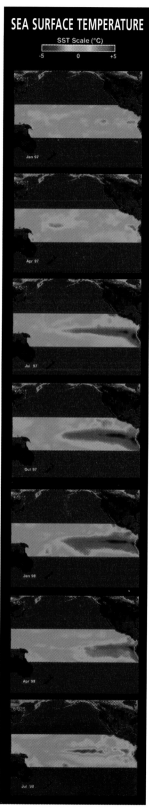

SEA SURFACE TEMPERATURE

SST Scale (°C)

-5   0   +5

Goddard Space Flight Center, NASA

NORMAL CONDITIONS

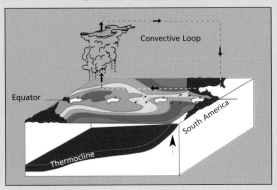

EL NIÑO CONDITIONS

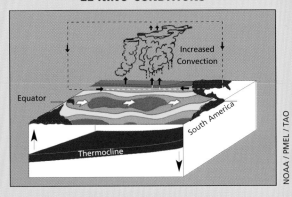

NOAA / PMEL / TAO

During El Niño years, warm water (in red) travels east, bringing with it moist air that drenches the west coasts of South America and North America. Cold ocean water below the thermocline (the layer separating warm and cold ocean water) pushes closer to the surface in the west, while in the east the warm pool forces cold water (in blue and green) farther down below the surface.

this cold water. As the cold water rises to the surface, great schools of fish swim to the area attracted by the abundance of food. Likewise, birds that feed on the fish flock to the region. This upward movement of the ocean water, called upwelling, creates some of the world's most productive fishing grounds. Fish are particularly abundant in the coastal waters off Peru.

The air is cool and dry along the coast where the ocean upwelling occurs. This prevents big rain clouds from forming; therefore, little rain falls on the arid coastal areas of Peru and Ecuador.

Pushed by the winds, a large pool of warm water collects along the coastal regions of Australia and Indonesia. In this western corner of the Pacific, the top layer of warm water goes down 400 feet (122 m) or more. As a result, the cold, nutrient-rich water is much deeper there and does not come to the surface as it does along the coasts of Peru and Ecuador. Warm, moist air above this warm ocean pool rises and forms clouds, which provide a good supply of rain to the coastal regions of the west Pacific.

In the normal sequence of events, the air above the Pacific Ocean moves in a clockwise motion from east to west at the lower levels and from west to east in the upper atmosphere. This cycle is called the Walker Circulation, named for Sir Gilbert Walker.

In most years the westward-blowing trade winds trap the warm pool of ocean water along the coasts of Australia and Indonesia. During an El Niño, however, the winds relax. When this happens, the warm pool of water is able to flow downhill across the ocean eastward toward Peru and Ecuador. Sometimes the winds actually reverse direction and speed the warm pool on its journey east. As it is pushed along, the warm pool forms a huge underwater mass called a Kelvin Wave. This giant wave—traveling east at five knots—covers 125 miles (200 km) a day. Spread out across

WALKER CIRCULATION

EL NIÑO SOUTHERN OSCILLATION (ENSO)

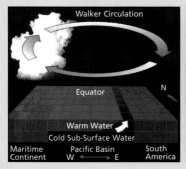
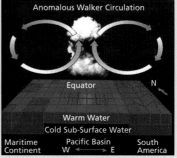

Goddard Space Flight Center, NASA

WALKER CIRCULATION

When there is no El Niño (pictured above, left), the top layer of warm ocean water is pushed by trade winds from South America westward to Australia along the equator. Cold waters well up to the surface along the Peruvian coast. The warm, moist air above Australia forms clouds that bring rain to the region. The colder air along South America prevents clouds from forming. As a result the coastal regions there are arid.

EL NIÑO SOUTHERN OSCILLATION (ENSO)

During years when there is an El Niño (pictured above, right), the trade winds relax and sometimes reverse direction. This causes the warm pool of water along the coast of Australia to flow back along the equator to South America. Warm, moist air above the pool of water brings rain to Peru and Ecuador, while in the western Pacific, drought conditions prevail.

6,000 (9,655 km) miles, the fiery El Niño has more force than one million atomic bombs—enough to heat a pool of ocean water the size of Canada.[11]

Sea level conditions are now reversed from normal; the sea level is higher along the South American coast and lower along the Australian coast. As the warm-water pool reaches the coasts of Peru and Ecuador, it raises the level and temperature of the ocean there and channels warm currents to the north and south. The ocean temperature in the eastern Pacific rises from one to four degrees Celsius or more during an El Niño event.

The warm waters heat the air above. The warmer the air, the more moisture it can hold. For each one-degree increase in temperature (on the Celsius scale), air can hold 6 percent more water vapor.[12] The water vapor forms clouds and eventually falls as rain or snow.

These changes in sea level and water and air temperatures have a dramatic effect on the weather. Because reversing winds and the warming of ocean water temperatures are both involved, scientists often refer to the ocean event as ENSO (El Niño-Southern Oscillation). Other scientists refer to it as a "warm event." Most people, however, simply call it El Niño.

two

REACHING ACROSS THE GLOBE

H ow did a pool of warm water in the Pacific Ocean near the equator result in an ice storm half a world away?

Under normal winter conditions in North America, westerly winds blow from the Pacific, collecting moisture from the Gulf of Mexico, the Great Lakes, and the Atlantic Ocean as they travel east. The warm, moist air collides with the cold air of northeastern United States and southeastern Canada, a prelude to precipitation.

In the weeks before the ice storm, El Niño disrupted this normal sequence of events. El Niño caused the warm trade winds to shift farther north. This shift in the wind pattern affected the northeastern United States as well as Quebec. Winds blowing from the south brought unseasonably warm, moist air to the region. The mid-Atlantic coast reached record high temperatures as a result.

The moist clouds of the warm front encountered the colder air from the north. Forced to ride over the cold air, the warm air was pushed high in the sky. Following the usual pattern, the moist air cooled, forming snow crystals that began to fall to the ground. But because the air close to earth was much warmer than normal, the crystals began to thaw as they fell to the ground. The droplets turned to freezing rain and covered the region with ice.

SETTING THE STAGE

El Niño itself does not create rain or snow, ice or drought. It merely sets the stage for unusual weather to occur. Droughts in Australia and Indonesia, for example, occur when El Niño moves the pool of warm water away from their shores and to the east. Deprived of a major source of heat, the air over that area of the Pacific does not warm and form clouds as usual. The seasonal rains called monsoons don't fall, and the region is left high and dry.

Floods occur when air saturated with warm moisture cools and falls in torrents. During El Niño, a giant cloud of humid air is heated as it floats above the warm pool of ocean water heading east. The air masses within the cloud cool as the warm air rises, forming water droplets and ice crystals and causing winds. Violent weather follows: heavy rains, wind gusts, and thunderstorms. The air above El Niño's huge pool of warm water holds so much moisture that it can rain in nearby regions for months on end. The

El Niño generates torrential rains and violent thunder-storms along the western coasts of South America and North America.

rains are frequently accompanied by huge thunderstorms, fueled by the heat given off by El Niño's warm-water pool.

The force of the storms send heated air and moisture 50,000 feet (15,240 m) or more into the atmosphere.[1] This disrupts air currents, sending warm winds farther north than normal and altering the jet stream's course. The jet stream is a narrow corridor 25,000 to 35,000 feet (7,620 to 10,700 m) above the earth along which high-speed winds blow from west to east. It affects and steers weather all along its course, from North America to Africa and Asia.

Bolts of lightning from the thunderstorms can touch off raging wildfires. In areas bone-dry from drought, like Indonesia and Brazil, such fires can ravage huge tracts of land. El Niño conditions also affect the formation of tornadoes. These storms are likely to form in the U.S. Southern Plains when cold, dry air from Canada meets warm, moist air from the Gulf of Mexico. Because the air in Canada is usually warmer when El Niño is around, the clash of air masses doesn't occur as it usually does. Thus, in El Niño years there are far fewer tornadoes in this area—known as Tornado Alley because of the frequency of storms there. Tornadoes are also less likely to occur in Louisiana, Arkansas, and Iowa.

Hurricanes also lose their punch in the face of El Niño. During El Niño years, tropical winds from the west blow the tops off hurricanes forming in the Atlantic before they have a chance to build up strength.

NOT THE ONLY ONE TO BLAME

Other factors—such as local weather conditions or people's behavior—can make El Niño's impact worse, turning a difficult situation into a disaster. The collapse of Peru's anchovy industry after the 1972–1973 El Niño is a good example. El Niño brought warm waters to Peru's shores. The anchovies normally fed on nutrients grown in the cold

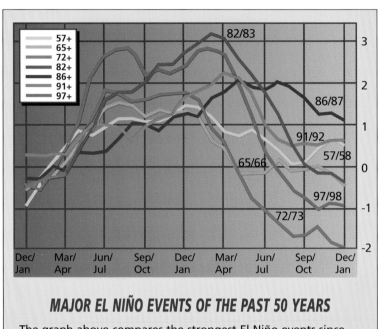

Source: NOAA/CIRES Climate Diagnostic Center

MAJOR EL NIÑO EVENTS OF THE PAST 50 YEARS

The graph above compares the strongest El Niño events since 1950. The lines show conditions of each event in relation to normal (represented by 0).

waters brought to the surface during upwelling. But because of El Niño, the upwelling was suppressed. Deprived of nutrients, some fish swam to more productive grounds. The fish that remained lived off their body fat. They became leaner and produced far fewer eggs than normal. As a result, baby fish were not being born to take the place of the adult fish caught by fishermen.

Peru's commercial fishermen became the second player in the tragedy. Unmindful of El Niño, the fishermen continued to haul in all the fish they could catch. During 1972 and 1973, the fishermen had huge catches. In fact, they caught almost all the fish off their shores. Normally, younger fish are small enough to swim through the holes in the nets. They survive to produce more fish for the next year's catch. But because of El Niño, there were few young fish in the area. The following year, fishermen came back to shore with empty nets. The industry collapsed. It took years for the fish stocks to recover.

THE COLD SISTER

Since 1900, El Niño has occurred 23 times, making its entrance every two to seven years. Usually, El Niño first appears in December to March and stays around from 12 to 18 months before finally disappearing. Often, but not always, the warm ocean event is followed by a twin that brings cold water to regions where El Niño had previously brought warm water. Called La Niña, meaning "the little girl," the cold event has appeared fifteen times in the last century. The latest, which surfaced in the central Pacific in the summer of 1998 and was expected to last through June 1999, touched off a record number of tornadoes in the United States during January 1999 and that same month dumped 63.5 inches (161 cm) of snow on Buffalo, New York.[2]

La Niña disrupts the jet stream, cools sea temperatures, and leads to extreme weather. It often produces weather patterns that are opposite of those provoked by El Niño. La Niña isn't as powerful as El Niño, but it still causes plenty of problems. Following as it does on the heels of El Niño, La Niña can deliver the finishing blows to regions still reeling from El Niño's destruction.

"We are more worried about La Niña than El Niño,"[3] said Victor Ramos, director of Environment and Natural

Resources in the Philippines, as his country prepared for flash floods expected to sweep over lands left barren by a thousand forest fires—El Niño's handiwork.

Like El Niño, La Niña is the result of Kelvin waves moving across the Pacific. When the huge warm pool of water that creates El Niño hits the South American coast, some of the water bounces back toward the west. These waves eventually bump against Asia, then slosh back again toward the east, but this time as a cold wave of water. La Niña carries the cool waters from deep under the ocean to the eastern Pacific.

The "little sister" cools the ocean water along its path by 1°C or more and marks the end of El Niño. In some years, the cool water merely cancels out El Niño's effects, returning the ocean to normal conditions. But in other years, for reasons not clear, La Niña cools the ocean enough to set off a reverse cycle of violent weather. During the 1998–1999 La Niña, water temperatures in some spots dropped more than 8°C in only two weeks.

As a result, droughts turn to heavy rains, balmy winters are followed by record-breaking snowfalls and freezing temperatures, and more hurricanes and tornadoes form along the Gulf Coast and the Atlantic.

Southern Atlantic states are often hit by violent thunderstorms. European explorers in 16th-century Florida marveled at the force of such storms, presumably during a long-past La Niña. René Goulaine de Laudonnière, a French explorer, wrote that "lightning in one instant" burned more than 500 acres (200 ha) of meadows near Jacksonville, Florida, in August 1587. The blaze, he said, continued for three days.[4] During a similar storm in north-

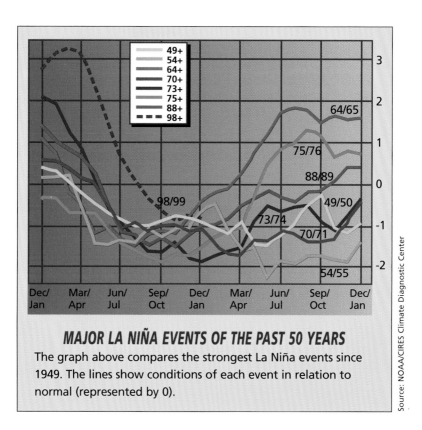

Source: NOAA/CIRES Climate Diagnostic Center

MAJOR LA NIÑA EVENTS OF THE PAST 50 YEARS

The graph above compares the strongest La Niña events since 1949. The lines show conditions of each event in relation to normal (represented by 0).

western Florida, Spanish explorer Nunez Cabeza de Vaca described how trees "were riven from top to bottom by bolts of lightning."[5]

While El Niño triggers weather that is the opposite of that normally seen in an area, La Niña tends to make usual weather conditions more extreme.

WHAT CAUSES EL NIÑO AND LA NIÑA?

Most scientists today agree on how El Niño and La Niña work. They have yet to determine what triggers the ocean events. Why do the trade winds suddenly relax, leading to El Niño's formation? Some researchers believe that tiny changes in heat from the sun may warm the ocean enough to get El Niño rolling. Another theory suggests that underwater volcanoes may be the source of heat that triggers El Niño.

Why does La Niña follow a warm event only some of the time? What determines whether it will be a major event or merely return things to normal? No one knows.

"It is fascinating that what happens in one area can affect the whole world," said Alan Strong of the National Environmental Satellite, Data, and Information Service, operated by the National Oceanic and Atmospheric Administration (NOAA). "As to why this happens, that's the question of the century. Scientists are trying to make order out of chaos."[6]

three

EL NIÑO THE CELEBRITY

Researcher Jacob Bjerknes showed strong links between El Niño and world weather in the 1960s, but many scientists remained unconvinced that the event had much force beyond Peru and Ecuador. Most of the public soon forgot El Niño after the 1957–1958 event.

The demand for Peruvian guano decreased when factories began to produce chemical fertilizers. With new products readily available, the periodic decline in guano birds held little interest among people living elsewhere. El Niño once again became a "Peru problem."

The El Niño of 1972–1973 changed all that.

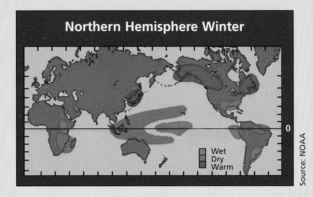

Northern Hemisphere Winter

Wet
Dry
Warm

Source: NOAA

FEAST AND FAMINE

El Niño's effects on weather can seriously reduce the amount of food produced around the world. The east coast of southern Africa, among other regions, has suffered from widespread famine as a result of droughts caused by El Niño. Floods in the west coast region of North America destroy crops as well. However, in countries along South America's western coast, farmers can grow more rice as a result of wetter-than-normal weather.

After farmers switched to chemical fertilizer, Peru's economy revolved around anchovies, the fish that swim in abundance off Peruvian shores. Beginning in 1953, plants in Peru processed anchovies into fish meal. Farmers in the United States and Canada used the fish meal as feed for pigs and chickens.

When a major El Niño struck in 1972–1973, Peru's fish industry collapsed. This reduced the amount of fish meal available to American farmers. They planted soybeans as a replacement on lands where they normally would have planted wheat. This reduced the amount of wheat available for the rest of the world. American farmers also had to cope with extremely dry conditions that season. The worst drought since the 1930s decimated the wheat harvest in Oklahoma, where farmers lost a quarter of their crops.

At the same time, dry conditions wiped out much of the grain raised in the Soviet Union, and droughts killed off crops in Central America, western and southern Africa, India, Indonesia, Australia, Ethiopia, and Brazil. Starvation threatened millions of people around the globe as nations dipped deep into emergency food supplies. For the first time since the end of World War II more than 25 years before, the world's supply of food was dangerously low.

The food crisis showed the world how El Niño—previously considered a local problem—could affect people around the world. Leaders held a World Food Conference in Rome in 1974 to address the food situation. Conferences sponsored by the United Nations on related topics followed.

Scientists began to focus on El Niño and its effect on weather worldwide. Television, newspapers, radio, and news magazines reported the global weather disasters and the famines that had claimed so many lives. For the first time El Niño became a household word.

Researchers set out to develop an early warning system so that people could prepare for El Niño.

THE EL NIÑO OF THE CENTURY

People have been monitoring the weather for centuries. The invention of the thermometer and the barometer in the 1600s allowed weather watchers to record temperature and air pressure, using a standard measure. Those weather reports could be shared with scientists around the world once the telegraph was invented in the 1800s.

With the help of weather reports collected by professionals and volunteers worldwide, scientists began to piece together the mosaic of global weather. They had no easy way of collecting data from the sea, however. Commercial ships provided readings of waters they crossed, but that left huge gaps in information. The data gathered from ships' logs covered only well-traveled routes at certain times of the year.

With the launch of weather satellites—beginning with the Television and Infra Red Observation Satellite (TIROS) in 1960—scientists finally had a way to view large expanses of the world's oceans. These orbiting weather reporters became invaluable to El Niño researchers as they focused their efforts on the Pacific Ocean. By studying water temperature, water currents, waves, and water pressure in the Pacific, the researchers hoped to detect slight changes that would signal the beginnings of an El Niño.

In early 1974, weather scientists predicted that a major El Niño would strike the following year. They found that currents were running faster than normal in the eastern Pacific Ocean and had observed warm water flowing from the north. The El Niño, however, never materialized.

The error reinforced some scientists' belief that ocean events couldn't be predicted months in advance. As a result they ignored the warning signs of an El Niño so severe that it would be called "The El Niño of the Century." Other scientists continued to believe in long-range forecasts, but they based their predictions on past El Niños.

But this was an El Niño like no other.

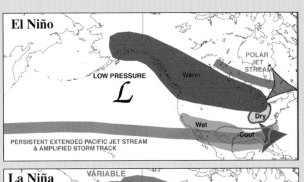

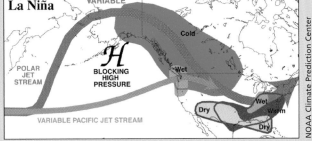

EL NIÑO, LA NIÑA & NORTH AMERICA'S WINTER WINDS

Top, El Niño pushes the polar jet stream farther north and shifts the Pacific jet stream to the south, producing warm winter weather in North America. *Bottom*, during La Niña, the warm, moist air of the Pacific jet stream shifts northward, where it collides with the cold air of the polar jet stream, which is pushed farther south than normal. This results in more tornadoes and hurricanes across the Midwest and along the Atlantic coast.

Early in 1982—the time of year when a typical El Niño usually starts to make itself known—conditions seemed normal in the Pacific. That year, forecasters happily predicted the American Midwest would have one of its largest harvests of corn ever. Scientists meeting in Seattle in June to discuss global weather issues found little to suggest that an El Niño was on the way.

Before an El Niño begins—or at least in the case of

warm events that had occurred during the past 30 years or so—easterly winds begin to blow with more force. This causes water to build up in the western Pacific. In the early months of 1982, there was no water buildup. The winds blowing from the east hadn't shown their muscle.

NAGGING DETAILS

Still, there were a few nagging details that suggested something was amiss. In May, weather satellites had detected a small rise in water temperature—less than 1°C—in the central Pacific. The next month the reports were even more puzzling. The data showed a sharp rise in air pressure off Australia and a much lower pressure in the atmosphere to the east. The readings were opposite those expected in the region.

Meteorologists reviewing the data at NOAA's Climate Analysis Center thought the figures were wrong.[1] During this time, a volcanic eruption was occurring, which scientists believed interfered with satellite transmissions. They thought that was the reason for the skewed data. To correct for the supposed errors, they programmed the computers to throw out figures that were too far from normal.[2]

NOAA meteorologist Eugene Rasmusson mentioned the strange data at the Seattle meeting, but no one seemed concerned enough to investigate further. A scientist attending a meeting of weather experts in Peru made a point of announcing that there would be no El Niño in 1982.[3]

In the Pacific, however, the easterly winds had stopped blowing and reversed direction. They began pushing a huge mountain of warm water across the Pacific. Deprived of the moist air that usually forms above the warm waters in the western Pacific, Australia, Indonesia, and New Guinea experienced a drought that June—another signal that El Niño had arrived.

Odd weather began to appear in the islands along the equator between Australia and South America. Tarawa, one of the small islands just north of the equator that make up the nation of Kiribati, had four times more rain than normal in July. Torrential rains spread east to other central Pacific islands as El Niño rolled on toward South America. In August, rain drenched Christmas Island (Kirimati) at the start of its dry season. Data from Australia showed the highest air pressure readings in 100 years. In Tahiti, pressures plunged to the lowest levels since the 1930s.

Things were looking pretty suspicious. NOAA issued a weather bulletin in September reporting some of the strange data. But scientists still didn't understand what the figures meant. The NOAA bulletin said only that the cause of the bizarre readings "is as yet unclear."[4]

The huge pool of heated water hit South America in October. The rains soon followed. They would continue for the next eight months. By November, water temperatures off the coasts of Ecuador and Peru had risen from the low 20s up to almost 26°C.

After it was all over, William Quinn, an oceanographer at Oregon State University, said scientists should have seen El Niño coming. "The signs were all there,"[5] he said. But by the time scientists recognized the warning signals, it was too late. El Niño had begun a rampage that would devastate the world.

The El Niño of 1982–1983 hit with a force that had

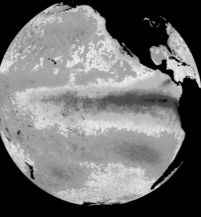

EL NIÑO'S TOLL—1982-83
TOTAL BILL: $13.09 billion

floods

Bolivia $300 million
Cuba $170 million
S. Brazil, N. Argentina,
  & E. Paraguay $3 billion
China $600 million
U.S. Gulf States $1.1 billion
Ecuador, N. Peru $650 million
Western Europe $200 million

drought / fires

S. Africa $1 billion
Iberian Peninsula, N. Africa
  $200 million
S. India, Sri Lanka $150 million
Philippines $450 million
Indonesia $500 million
Australia $2.5 billion
S. Peru, W. Bolivia $240 million
Mexico, Central Amer. $600 million

hurricanes

Tahiti $50 million
Hawaii $230 million

storms

U.S. Mtn. & Pac. States
  $1.1 billion
Middle East $50 million

not been seen in at least 100 years. Some scientists believe it was the worst on record. After reexamining its data in late fall, NOAA issued a new bulletin belatedly announcing an El Niño "of major proportions."[6] The warm event, NOAA predicted, would be as fierce "or possibly even exceed" the 1972 El Niño.[7]

Throughout the next few months, El Niño raged. By February 1983, it appeared that the event had spent its energy. Water temperatures dipped, winds and currents changed direction, and atmospheric pressures began to return to normal. NOAA issued a tentative bulletin that the event "may have passed its peak."[8]

It hadn't. A month later, waters off the coasts of Peru and Ecuador had reached the high 20s. Before El Niño was done, sea temperatures at the equator would register an incredible 32°C. Once again, currents of wind and water reversed themselves. El Niño was back with a vengeance.

WIDE SWATH OF DESTRUCTION

A look at countries around the world tells the story. El Niño left its mark of death, disease, and destruction on every continent but Antarctica. The nations bordering the Pacific fared among the worst. Six hundred lost their lives when floods demolished Ecuador and northern Peru at a cost of $650 million. More rain fell on Peru than ever recorded any place on earth. Eleven feet (3.3 m) of rain inundated lands that normally received only 6 inches (15 cm) of rainfall a year.[9] Southern Bolivia, Ecuador, and the Galapagos islands endured months of rain that destroyed 40,000 of the area's adobe homes. Boats became the only vehicles that could travel along the Pan American Highway and other major roads.[10]

On the southern side of Peru and in northern Bolivia, where the region's farms produce food for local use and for export, crops withered in the endless heat. An eight-month drought destroyed all but a tenth of Bolivia's potato crop. Raging floodwaters in southern Brazil, northern Argentina, and eastern Paraguay forced 600,000 people to evacuate, killed 170, and cost those nations $3 billion.[11]

Nations to the north suffered from scorching heat or constant downpours. A drought in Mexico and Central America cost inhabitants $600 million. Fifteen drowned in Cuba, where sugarcane rotted under water, and floods caused $170 million damage.[12]

Peru's fur seals were among El Niño's victims.

Blizzards, mud slides, floods, and tornadoes ravaged the western United States and Rocky Mountain region, causing damages of $1.1 billion. Forty-five people died, crops and cattle were destroyed, beaches eroded, and homes and businesses were lost. Along the Gulf of Mexico, flooding claimed 50 lives and caused another $1.1 billion in damages.[13]

The nations in the western Pacific endured weather-related disasters equally devastating. Uncontrolled brush-fires destroyed the homes of 8,000 in Australia. There had not been a summer so dry on the continent in the past 100 years.[14] The drought and fires left 71 dead; damages totaled $2.5 billion. Indonesia lost 340 people to starvation and suffered a half-billion-dollar loss due to a record drought. Wet weather in China was blamed for the deaths of 600 and losses of $600 million.[15]

Africa and Europe felt the deadly effects of El Niño as well. Snowstorms in the Middle East, where a desert climate usually prevails, caused $50 million damage and left 65 people dead. Southern African nations struggled with starvation and disease brought on by a severe drought. The cost of that crisis was estimated at $1 billion. A $200 million drought plagued northern Africa, Spain, and Portugal.[16]

In March 1983, a scientific expedition reported one of the strangest happenings associated with El Niño. The entire bird population of Christmas Island disappeared. Finding few fish in the warm waters surrounding the atoll, the birds flew in all directions in search of food. They left behind millions of starving chicks. Six months later, after El Niño had retreated, the birds began to return to their island home, but their offspring had all died by that time.

Similar conditions claimed the lives of a quarter of Peru's adult fur seal and sea lions. Their offspring, like the chicks, starved to death.

Worldwide, El Niño-related weather in 1982–1983 caused more than $13 billion in damages and was responsible for up to 2,000 deaths.[17] Scientists learned that the forces resulting from the ocean event were strong enough to slow down the rotation of the earth. By doing so, El Niño added one-fifth of a millisecond to several days in January 1983 during its rampage.[18]

NOT ALL BAD

Despite the devastation that trails the event, El Niño, like most things in nature, is not all bad. In 1982–1983, most of North America enjoyed the warmest winter in 25 years, thanks to El Niño's effect on the polar jet stream. El Niño pushed the cold wind current farther north, sparing northern United States and much of Canada from its icy blasts. Northerners saved an estimated $500 million in heating bills that winter as temperatures rose an average of nearly 3°C above normal.[19]

Lands along the Atlantic Ocean got a reprieve from hurricanes during El Niño's tour. Canadian fishermen netted a big salmon catch after the fish swam north to escape the warm waters El Niño had brought to the U.S. Northwest.

James O'Brien of the Center for Ocean-Atmosphere Prediction Studies at Florida State University decries El Niño's bad reputation. "It is a shame that almost everyone blames bad events such as floods, droughts, fish kills, and heat waves on El Niño,"[20] O'Brien said. He notes that the event brings fish to southern Ecuador waters, deposits shrimp on Peru's shores, and transports Peru's anchovies to Chile. In the southeastern United States, he continues, "El Niño is wonderful. It brings gentle, extra rain in autumn and winter, which in the dry season suppresses forest fires from Arizona to Florida."[21]

four

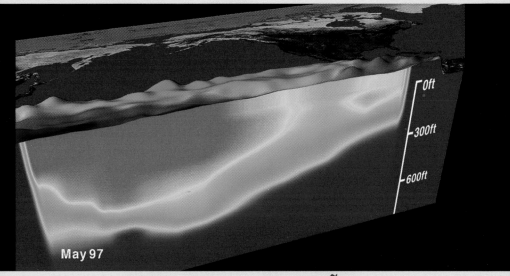

May 97

0ft

300ft

600ft

TRACKING EL NIÑO

T he El Niño of 1982–1983 left weather scientists
shaking their heads in awe. "When you see what
can happen, how a quirk in the wind can do all
this, it's humbling," said Alan Strong, a scientist at
NOAA's National Environmental Satellite, Data, and
Information Service. "It reminds us what Mother Nature
can do."[1]

Weather forecasting has come a long way since the
1960s, when scientists first began to use computers to plot
weather maps instead of plotting them by hand. Since
then, computers have been put to use collecting and ana-
lyzing all kinds of weather data and predicting weather

conditions. Weather satellites, first launched in the 1960s, track weather conditions around the globe. Sensors aboard the satellites can detect air and ocean temperatures, track clouds, measure moisture in the air, gauge sea levels, and spot hurricanes, tornadoes, thunderstorms, and blizzards as they form. Some take pictures of Earth every few minutes and beam them to weather stations, where they are used to track storms and other weather and ocean events. Buoys and commercial ships report on currents below the ocean surface and other readings that satellites can't detect from space.

The equipment, however, is only as good as the experts trained to read and interpret its findings. During the 1982–1983 El Niño, scientists ignored the warnings in the data from the Pacific. Information from weather satellites, data from buoys, and reports from commercial ships all showed signs that a huge El Niño was developing. But the results were too unexpected to be believed. The experts assumed the equipment wasn't working properly.

After the 1982–1983 El Niño disaster, government leaders as well as scientists knew they had to develop a better system to monitor ocean events. To do that, the United Nations World Meteorological Organization set up the Tropical Ocean-Global Atmosphere (TOGA) program in 1985. TOGA used an arsenal of tools: satellites, ships, tide gauges, and moored and drifting buoys. This equipment provided continuous reports on the temperature of the ocean water, the movements and strength of the wind, sea level, air and water pressure, and ocean currents.

The El Niño of 1986–1987 gave scientists a chance to test the new TOGA system. One thousand experts from

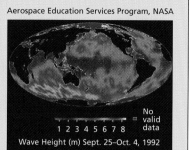

Aerospace Education Services Program, NASA

No valid data

1 2 3 4 5 6 7 8

Wave Height (m) Sept. 25–Oct. 4, 1992

SPACE-AGE DETECTIVES

Jet Propulsion Laboratory, NASA

The Topex/Poseidon satellite, pictured in an artist's drawing above left, measures the height of the sea's surface. Above right, an image created from data transmitted from Topex; the red areas indicate high sea levels, while the blues and greens pinpoint lower sea levels. The reading was taken in the fall of 1992.

around the world monitored the event's progress. They used data from seven satellites, fourteen ships, and a network of buoys that radioed information to researchers via satellite. For the first time, scientists were able to track an El Niño from beginning to end. What they learned would help the world prepare for another "storm of the century" that surpassed the 1982 event in destruction and power.

FORECASTS: A TRICKY PROPOSITION

A region's prevailing weather conditions, known as the climate, follow a typical pattern. For example, winter in the Northeast usually brings cold and snow, while summer is

View of a typhoon taken from a weather satellite

the time to grow crops and bask in the sun. In warmer climates, rainy seasons and dry spells occur regularly. People know what type of weather to expect in a particular region during each season.

The weather—the atmospheric conditions on a certain day—may or may not follow the general pattern of the climate in that area. Although most days in winter are cold in the Northeast, occasionally temperatures will hit 10°C or even 15.5°C in January or February. Though snow falls every winter, it doesn't storm every day, and some winters have heavy snowfall while in other years there is little snow. "Climate," says social scientist Michael Glantz of the National Center for Atmospheric Research in Boulder, Colorado, "is what you expect. Weather is what you get."[2]

Predicting the weather has always been a tricky proposition. While some regions have the same conditions day after day, most areas experience a constantly changing weather picture. Ants Leetmaa of the National Weather

Service's Climate Prediction Center refers to the changing climate's "irregular heartbeats."[3] Humorist Mark Twain once remarked about New England weather, "In the spring I have counted 136 different kinds of weather inside of 24 hours."[4]

Meteorologists admit that even with the most advanced technology available today, they cannot predict with absolute certainty what the weather will be on a particular day. Conditions can change quickly, without notice. For example, a storm headed right for a particular region can suddenly shift course, blown out to sea by a strong wind or stalled 90 miles (145 km) to the west by a change in air pressure. Yesterday's forecast of a 90 percent chance of rain is discarded. A new forecast predicts sunny weather for the next day or so.

Developing a forecast months in advance is even trickier. If scientists can tell when a major El Niño is likely to occur, nations have a chance to prepare for the worst. Such predictions, however, may not always come true. Scientists are faced with a dilemma. Should they issue an alert when they are not sure an event will develop? Or it is better to wait and be sure before notifying the public? Both courses can be dangerous.

The 1973 prediction of a major El Niño that never developed led people—scientists included—to doubt predictions that followed. During the 1997–1998 El Niño, forecasters expected a severe drought in southern Africa. Taking the warning to heart, government officials set aside food for the expected shortage. As it turned out, the region had a normal rainfall. Some crops were in short supply, however, because farmers, not wanting to risk losing their harvest to El Niño, had decided to reduce their plantings that year.

Wild speculation from a public too quick to blame El Niño for every storm does not help the situation. The public, says scientist Glantz in his book, *Currents of Change*, "expects perfection."[5] But weather predictions, especially long-range ones, aren't perfect and probably won't be for many years to come. Forecasters state the chances that an event will occur; good forecasts are right more often than not, but they also can be wrong. "A forecast is a forecast," notes Glantz. "It's not a guarantee."[6]

Gilbert Walker warned scientists in the 1930s against making surefire predictions. If they turn out to be wrong, he noted, people won't believe the forecast next time. "It is the occasional failures of a government department which are remembered," he said. "It is better to admit your limitations, and only speak when you can do so with some safety than to issue predictions when they are little more than guesses."[7]

EVERY EL NIÑO DIFFERENT

With all the advanced equipment now available, why is it still so difficult to predict an El Niño? All El Niños follow certain patterns, but each is unique. There is no "normal" El Niño. The past offers valuable information on what to expect, but each El Niño has its own particular twists. The 1982–1983 El Niño, for example, surprised most scientists because it didn't begin early in the year as an El Niño typically would do. Before the event, scientists detected no overly strong easterly winds—a signal that preceded the major El Niños in 1957–1958 and 1972–1973.

Researchers deploy an ATLAS mooring, part of the monitoring system used to measure surface winds; air, sea surface, and subsurface temperatures; and relative humidity in the Pacific Ocean along the equator.

Even when they accurately predict a major El Niño, scientists still can't say for sure how it will affect a particular area on a certain day. Any number of things can influence El Niño's behavior, from local weather conditions to global warming. You might be able to blame El Niño for a wetter than normal summer, but you can't really say whether it was El Niño that caused the cloudburst on your birthday.

December 1994

NOAA / PMEL / TAO

TOGA MONITORING SYSTEM IN THE PACIFIC

The map above details the location of the buoys, ships, and other segments of the complex TOGA (Tropical Ocean Global Atmosphere) monitoring system that measures conditions in the Pacific Ocean.

That rain might have happened anyway due to a front that suddenly moved into the area that day.

With each new El Niño, scientists learn more about the event and how it works. Using this information from events in the past, scientists improve their mathematical models of El Niño. These models show how El Niño might behave under various conditions. The models can also be used to help predict the next El Niño.

In early 1986, Stephen Zebiak and Mark Cane, working with computer models at Columbia University's Lamont-Doherty Earth Observatory, became the first to accurately predict an El Niño several months in advance. Their success not only proved that a long-term forecast was possible; it also saved lives.

In one instance, Ethiopian farmers, warned by the forecast, planted a much bigger crop than usual during the short rainy season from mid-February to mid-May. Without the warning, they would have waited to do most of their planting during the rainy season that normally occurred later in the year. By the middle of 1986, a moderate El Niño had developed. A severe drought hit Ethiopia. But because of the large crops planted in the spring, the country was spared a famine. Not one person died from hunger, despite El Niño's appearance.[8]

five

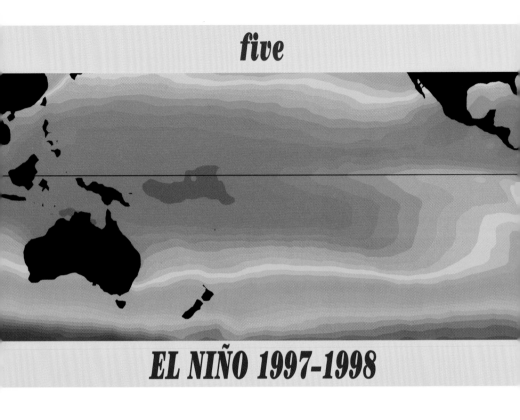

EL NIÑO 1997-1998

*T*he success of Zebiak and Cane encouraged scientists to work even harder to track and predict future El Niños. Forecasters correctly predicted an El Niño in 1991, but they mistakenly predicted its effects would fade by the end of the year. Droughts unexpectedly continued into 1992. Temperatures in the western Pacific continued to be warmer than normal until early in 1995. Some scientists believed that three short El Niños had occurred, while others thought the warmer waters were due to one long El Niño.

Again El Niño had deviated from the expected course and had forced scientists to concede that the ocean event still held mysteries as yet unsolved. Still, weather forecasts relating to El Niño helped nations prepare for droughts and floods in the early 1990s. Even with the failure of some long-range forecasts, the monitoring system in the Pacific provided accurate information on El Niño's progress once it appeared.

TOGA, the ocean and satellite tracking system, was enhanced in 1994 by a new series of sophisticated satellites and buoys that constantly monitor the air and the sea along the equator in the Pacific Ocean. The Tropical Atmosphere Ocean (TAO) array, a string of 70 buoys equipped with complex instruments, beams data via satellite from the Pacific to scientists at weather stations throughout the world. Computers compile the information for up-to-the-minute reports on everything from ocean color to wind direction. The buoys, which cost about $50,000 each, cover an 8,000-mile (12,874-km) stretch of ocean along the equator from New Guinea to Panama.[1]

The TOPEX/Poseidon satellite, launched in 1992, has mapped much of the ocean's surface, using radar to measure sea level. The high spots indicate warm areas, because warm water expands. Cold water lies in areas where the sea level is low.

Instruments aboard other satellites measure the amount of moisture in the upper air. Large amounts of water vapor in the atmosphere where it is not normally found is a clue that El Niño has arrived.

In early 1997, scientists reviewing this constant flow of information saw a familiar pattern. It suggested that an El Niño was on its way. Some computer models revealed the same thing, although the Cane-Zebiak model was predicting a weak La Niña.[2] Most scientists believed the El Niño would be a weak one.

Once El Niño developed in March 1997, scientists revised their forecast. New data revealed the event would be much stronger than originally anticipated. Based on these findings, NOAA told the world in April 1997 that a major El Niño was developing in the western Pacific. It marked the first time scientists had correctly predicted the arrival of a major El Niño more than six months before its effects were felt. The following month, weather scientists in Australia and Japan issued their own warnings.

By summer, scientists were forecasting El Niño's expected effects on the world's nations. They used data from TAO and images from TOPEX/Poseidon to track El Niño every step of the way as it followed its roller-coaster course across the Pacific. The world was well prepared for this El Niño, and as a result, thousands of lives were saved.

"With this event," said NOAA scientist Dr. Ants Leetmaa, "we were light years ahead of the last major El Niño."[3]

Data from buoys, satellites, and computer readings kept scientists well informed of El Niño's progress in 1997–1998. The images on the facing page show the changes El Niño produced in the height of the sea surface as measured by the TOPEX/ Poseidon satellite (A), the underwater temperature as measured by TAO buoys (B), and the height and temperature of the surface of the ocean (C). The height is represented by the peaks in the latter graphic; temperature is portrayed by color, red for hot and blue for cold. Beginning in January 1997, the images show El Niño's influence and that of La Niña, which appeared in mid-1998.

SEA SURFACE HEIGHT (A)

SSH Scale (cm)
-30   0   +30

Jan 97

Apr 97

Jul 97

Oct 97

Jun 98

Apr 98

Jul 98

UNDERWATER TEMPERATURE (B)

SST Scale (°C)
20   28

Jan 97

Apr 97

Jul 97

Oct 97

Jan 98

Apr 98

Jul 98

HEIGHT & TEMPERATURE (C)

SST Scale (°C)
-5   0   +5

Jan 97

Apr 97

Jul 97

Oct 97

Jan 98

Apr 98

Jul 98

Goddard Space Flight Center, NASA

El Niño disrupted weather patterns, causing violent storms that led to billions of dollars of damage throughout the world.

WILD ROMP

The warnings, however, couldn't prevent El Niño-related disasters around the world: flooding in Peru, Ecuador, Mexico, and the American West Coast; drought and wild-fires in Indonesia, Australia, and Brazil; tornadoes and floods in the American Southeast; a devastating ice storm in the American Northeast; and droughts and starvation in southern Africa.

In May 1997, NOAA scientists predicted this would be the worst El Niño in 150 years. It soon became apparent that the event would live up to its billing. Peru, as always, was among the first to feel the fury of El Niño weather. Violent weather began there in late 1997 and continued for months. In January and February 1998, rains soaked the normally dry regions of the coastal plains. The floods and mud slides that resulted claimed the lives of 300 people. Thousands more lost their homes. After weeks of rain, a raging river overflowed its banks and covered the entire town of Chato Chico. Mud slides buried other towns. In Mampuesto, Peru, skeletons and caskets floated down the streets when a dam burst and undermined a local cemetery.[4]

The Pan American Highway looked like the scene of a battlefield. Huge holes in the road and collapsed bridges cut off all travel except by foot and inner tube. By March 1998, rains had washed away 59 bridges in Peru, damaged 28 more, destroyed 530 (852 km) miles of road, and left 3,880 miles (6,244 km) of highway pitted with holes and ruts.[5]

Across the Pacific, Australia coped with the worst drought on record. Forest fires, made worse by the dry conditions, burned out of control, destroying homes and wildlife habitat.

For more than a year, El Niño brought record-breaking weather to the nations along its deadly path—from rainfall and droughts to hurricanes and typhoons. Here is a partial list of the weather records and near-records broken during El Niño's wild romp:

- Rainfall in southern California was 231 percent above normal during 1997–1998. Erosion, caused by wind-whipped waves and driving rains, undermined cliffs, causing luxury beach houses to fall into the sea. Sand washed away from beaches. By July 1998, Las Tunas State Beach in Malibu had "virtually disappeared."[6]

An image taken from a satellite shows Hurricane Nora as it whirls off the coast of Baja California, Mexico. El Niño's warm waters kept Nora closer to Mexico's coast than would have occurred otherwise, according to scientists.

- In Hilo, Hawaii, known as the wettest city in America with an average of 10 $\frac{1}{2}$ feet (3.2 m) of rain a year, people's water supplies had to be restricted when El Niño conditions brought a severe drought. By February, only $\frac{1}{2}$ inch (1.26 cm) of rain had fallen on the region, where 15 inches (38 cm) is normal.[7]
- Eighty-six people died in floods in Kenya, the worst flood-related death toll in that nation for many years.[8]
- The worst tornado in Florida's history swept through that state in February 1998 at winds of up to 250 miles (400 km) per hour, leaving behind smashed homes and businesses and 40 dead.[9]
- The worst drought in 50 years in Papua New Guinea threatened one million people with starvation. A famine in Indonesia killed 650 after drought destroyed that nation's crops.[10]

- Record rainfall in Somalia caused the Juba River to overflow and sent 300,000 people to tent cities in the hills.[11]
- The World Wide Fund for Nature named 1997 "the year the world caught fire." A record number of tropical forests burned, causing respiratory problems and deaths in people and threatening the lives and habitats of a wide range of animals. Indonesian fires spread pollution and choking smoke across six Southeast Asian nations. Wildfires in the Amazon rain forest—spreading rapidly during an El Niño drought—caused similar problems in the Western Hemisphere. Thick smoke from the fires carried all the way to Texas.[12]
- Guam became the windiest spot on the globe when a typhoon ripped across the island nation, causing $100 million in damage in December 1997.[13]
- Two hundred people died when a typhoon struck China and Taiwan, the region's worst in 10 years.[14]

According to Ants Leetmaa, director of NWS Climate Prediction Center, El Niño was linked to all 18 of the federally declared disasters that struck the United States during the winter of 1997–1998.[15]

Some areas benefited from El Niño's influence. Heavy rains cleansed the air of smog in southern California. Joggers breathed fresher air, and mountains on the outskirts of Los Angeles—previously invisible in the haze—appeared on the horizon. Along the Atlantic coast, the hurricane season was the gentlest in years. Only one hurricane hit land, Danny, in mid-July 1998. That storm caused $100 million damage in the Carolinas and cost nine people their lives.

The warmer winter saved Northerners in the United States billions in fuel costs. As in the past, rains turned desert areas into new planting grounds.

"EL MEANO"

Merchants had mixed feelings about El Niño. Some benefited from the ocean event; others faced heavy losses. Companies that depend on snow experienced downturns in business during the mild winter of 1997–1998 in the U.S. Northeast. Ski operators in Maine dubbed the ocean event "El Meano."[16] Outdoor sports equipment manufacturers blamed El Niño for a dip in their sales. American farmers, too, took a hit when rains or droughts damaged crops from Florida to California. Caesar salad practically disappeared from the nation's menus after heavy rains in February 1998 destroyed California's romaine lettuce crop. Western dairy farmers lost 6,500 cows when the animals became stuck in the mud and starved to death. In New Hampshire and Maine, apple growers reported a meager harvest because of the warm winter and wet spring.[17]

Retailers thanked El Niño when warm weather brought shoppers to malls in February 1998. They cursed the ocean event when heavy rains kept shoppers home. Exterminators had an unexpected boost in their business when moist, mild weather produced swarms of killer bees, carpenter ants, ticks, rats, and fleas. The bug bonanza, however, also boosted the number of cases of encephalitis, a brain disease, and malaria, both spread by mosquitoes; Lyme disease, carried by ticks; and other diseases.

When El Niño finally ended in the summer of 1998, scientists blamed its direct influence for the deaths of at

El Niño conditions brought droughts to Brazil, Australia, Africa, and Asia that killed livestock and destroyed crops. Heavy rains in California killed 6,500 head of cattle when the animals became mired in mud.

least 2,100 people and damages of $33 billion.[18] Weather-related disasters that year—many of which could be traced directly or indirectly to El Niño or La Niña—racked up an incredible toll in death and destruction. According to the Worldwatch Institute, violent weather killed 32,000 people and caused $89 billion in losses from January to November 1998. Three hundred million fled their homes, and countless others suffered injuries and ill health because of storms and other weather disasters.[19]

Wildlife suffered as well. Orangutans fought for survival after fires in Indonesia burned their food and their habitat; elephants in Kenya suffered in blazes there; and baby sea lions died in Chile when their mothers abandoned

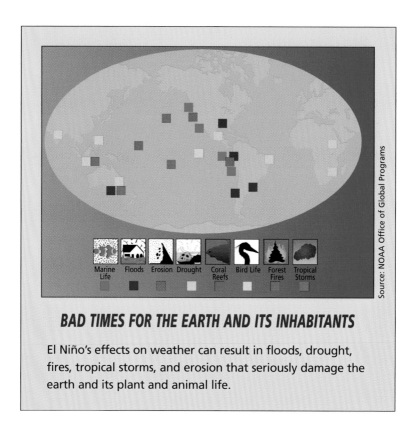

Source: NOAA Office of Global Programs

Marine Life	Floods	Erosion	Drought	Coral Reefs	Bird Life	Forest Fires	Tropical Storms

BAD TIMES FOR THE EARTH AND ITS INHABITANTS

El Niño's effects on weather can result in floods, drought, fires, tropical storms, and erosion that seriously damage the earth and its plant and animal life.

them to search for food in cooler waters. Warm waters off San Diego shores decimated kelp forests and sea urchins.

This El Niño hit the top of the scale; its ferocity surpassed even the "storm of the century" in 1982–1983. "It was," as Michael McPhaden, a director at NOAA's Pacific Marine Environmental Laboratory in Seattle, said, "in the category of a super El Niño."[20]

LA NIÑA, 1998-1999

By May 1998, scientists were announcing the arrival of La Niña, El Niño's "cold sister." Under La Niña's influence, the Pacific Northwest had a wetter winter than usual, southern California had less rain, and northern areas had record cold and snow. A drought in America's southern plains cost farmers and others an estimated $7 billion, according to NOAA.[21] The northern snow cheered skiers but caused avalanches that claimed several lives. A January 2, 1999, blizzard buried Chicago under 18.6 inches (47 cm) of snow, and 13 days later extremely heavy snowfall forced Detroit to declare a state of emergency.[22]

La Niña's cold spawned nine hurricanes and five tropical storms in the Atlantic in 1998, the deadliest hurricane season in two centuries. Hurricane Mitch, the fiercest, formed October 22 and swept across the Caribbean, blowing gusts of wind clocked at more than 200 miles (320 km) per hour. The death toll from Mitch's torrential rains, floods, and the mud slides that resulted topped 10,000 in

The chart at right compares the sea surface temperatures during the 1982–1983 and the 1997–1998 ENSO events. Red indicates warm water; blue is cold water.

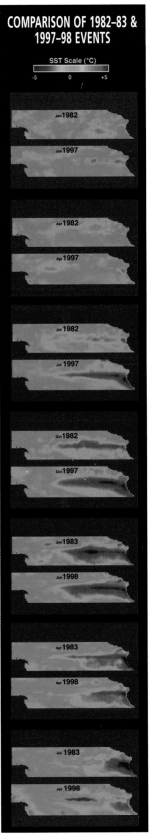

COMPARISON OF 1982–83 & 1997–98 EVENTS

SST Scale (°C)

-5   0   +5

Jan 1982
Jan 1997
Apr 1982
Apr 1997
Jul 1982
Jul 1997
Oct 1982
Oct 1997
Jan 1983
Jan 1998
Apr 1983
Apr 1998
Jul 1983
Jul 1998

Goddard Space Flight Center, NASA

Honduras and Nicaragua. It was the third-deadliest Atlantic hurricane on record.

More than 100 tornadoes—almost double the previous record—struck the United States in January 1999, mostly in the South. A major ice storm hit Maryland on January 15, 1999, turning off the electricity in 43,000 homes and closing schools and offices. On January 14, 1999, the temperature in Alagash, Maine, dipped to a frigid -30.5°C, the coldest ever recorded there.[23]

South Africa's mild weather during El Niño abruptly changed to wild weather under La Niña's sway. President Nelson Mandela barely escaped injury when the pharmacy building where he was shopping collapsed during a vicious tornado in December 1998. The president's guards protected him with their bodies. The twister leveled the town and killed 18 people.[24]

Wet weather was expected in northern Australia, Indonesia, and India. It was only the third time in 20 years that a La Niña had developed. It followed El Niño in 1987 and 1995. Because it is rarer and usually less intense than El Niño, La Niña is more difficult to predict. Scientists watching the phenomenon were hoping to learn more about the cold event.

AIDING AND ABETTING: PEOPLE'S ROLE IN DISASTERS

El Niño gets the blame for weather disasters, but too often people's actions and local weather conditions are as much at fault as El Niño. For weeks in the fall of 1997, a deadly haze covered much of Southeast Asia. An El Niño drought was blamed for spreading fires in the region. The smoky haze, however, originated when logging companies set fire

A tornado makes its deadly sweep across the land. In January 1999 more than 100 tornadoes struck the United States, nearly double previous records. The conditions that produced the storms were blamed on La Niña, which brought unusually cold air to the eastern two-thirds of the United States. The cold air collided with warm, humid air from the Gulf of Mexico, creating an ideal environment for tornadoes.

to thousands of acres of land they wanted to clear for farming. Normally monsoons put out the fires and clear the air. But because of El Niño, the region was extremely dry. The fires spread rapidly, and their smoke mixed with dust from the dry land and emissions from factory smokestacks, cars, and trucks.

The smog, laden with pollution, spread over six countries, covering 2,000 miles (3,218 km) and affecting up to 70 million people. The haze was so thick in some areas that people wore surgical masks to keep out the choking smoke. In downtown areas, tall buildings appeared as "ghostly shadows."[25] Local newspapers urged readers to stay indoors, wash frequently, and stop smoking. In parts of Malaysia, the air pollution count reached a record 839 ppm of sulfur dioxide, carbon dioxide, nitrous oxide, lead, and dust. Levels between 301 and 500 are considered hazardous.[26]

More than $1 billion damage resulted from the effects of the haze. Hundreds of people died, and tens of thousands became ill when pollutants in the air triggered asthma attacks and other respiratory problems. Doctors were concerned that exposure to the smog would lead to lung cancer and other ills. Inhaling the smoky air, experts said, was like smoking five packs of cigarettes a day.[27] A plane crash in Sumatra on September 26 killed all 234 people aboard when smog clouded the pilot's view of the runway.

Other disasters have also had an assist from people. In California, builders construct luxury houses atop cliffs or along beaches where storm-whipped waves wash them out to sea. Flooded rivers have swept away entire towns only to have residents return to build again. In overcrowded regions, people often have no choice but to live along riverbanks or wetlands.

In China loggers and farmers cut down most of the trees along the Yangtze River, which caused it to overflow much more quickly during heavy rains. The resultant floodwaters killed 3,700 and forced almost a quarter million people to seek shelter on higher ground.[28]

Scientists wonder whether global warming—the gradual increase in the earth's atmospheric temperature caused by burning fossil fuels—will affect El Niño. Statistics collected over the past century seem to show that the event is

Smoke from uncontrolled fires obscures the island of Borneo. The fires blazed for months, sending choking smog throughout Southeast Asia and endangering the health of thousands. The image is from the NOAA-14 Polar Orbiting Environmental Satellite.

getting stronger. Four of the ten worst events, including the two most destructive events ever recorded, occurred after 1980. Some scientists believe that the apparent increase in major El Niños may be due to global warming. If that is the case, they say, the results could be disastrous for the earth.

"If we have these very hot years more often as a result of increased global warming," said Doug Inkley, a scientist with the National Wildlife Federation, "the catastrophic impact on people and wildlife we've seen with this year's El Niño could be just the beginning."[29]

Even if the increase isn't connected to global warming, environmentalists say El Niño gives a good preview of what to expect if the earth continues to heat up. "El Niño is like a window in which we can look at the future if global warming continues—it's representative of a very hot year,"[30] said Inkley.

six

FORECAST FOR THE FUTURE

S urveying the damage from the most recent El Niño and La Niña, one might wonder whether scientists' predictions of the events served any purpose. Without the early warnings, however, the damage and loss of life would have been far greater.

El Niño hit the U.S. West Coast with more force in 1997 than it had during the 1982–1983 event. San Diego County received almost twice as much rain as it usually does. But the damage was less because residents were ready for the event, and local conditions were more favorable. People cleared storm drains and took steps to reduce coastal erosion. In addition, tides in the area weren't as high during the most recent event. The heavy rains cost the county about $10 million; no lives were lost. A similar storm in 1993, however, left 24 people dead and caused $193 million in damage.[1]

Between storms, residents and work crews piled sandbags along swollen rivers and coastal areas, braced piers and other structures, and cleared mud and other debris from beaches. Emergency workers were ready to evacuate residents to shelters.

Preparing for the worst, the International Red Cross asked for $167 million in January 1998 to deal with El Niño-related disasters expected that year. The organization estimated that more than 10 million people in 56 countries would need help because of El Niño. Because of the early warnings, the group had provisions and workers ready when people needed them. Other emergency workers prepared evacuation routes and equipped themselves to battle expected fires and floods.[2]

"We did a really good job of preparing for it this time," said Kathy Tice, a spokesperson for the Emergency Operations Center in Monterey County.[3]

TAKING EL NIÑO SERIOUSLY

Governments are taking El Niño seriously. Latin American nations budgeted hundreds of millions of dollars to deal with El Niño-related disasters in 1997–1998. Peru borrowed $250 million from the World Bank and other lenders to take precautions against El Niño's force. Government agencies used the money to reinforce bridges and roads and help families displaced by weather disasters related to El Niño.[4] The United States is contributing $18 million toward a research center in New York devoted to El Niño studies. Nations from around the globe are expected to participate.[5]

American fisheries officials sharply limited fishermen's catches in Pacific fishing grounds after El Niño-heated waters reduced the fish population there. The restrictions, which applied to cod, sole, rock fish, and other species, prevented fishermen from wiping out the fish stock.[6]

In Mexico, residents along the riverbanks heeded warnings that Hurricane Rick was headed their way in November 1997 and fled to higher ground. As a result, no one was killed by the hurricane, which was linked to El Niño conditions.[7] The month before, more than 100 people died in Mexico when floodwaters swept them away during Hurricane Pauline. Residents had had no warning of that storm.

Before El Niño hit, Peruvian workers built storm drains to divert raging waters and set aside emergency supplies for victims of disasters. Their actions saved hundreds of lives.[8]

Accurate forecasts can help people adapt to the weather changes El Niño brings and even use them to their advantage. Farmers in northern Peru moved cattle to new grasslands created when rains transformed deserts into fertile ground. They also used the rain-soaked deserts to plant rice and beans. Fishermen prepared to catch shrimp instead of anchovies along Peruvian shores. In Australia, where drought was expected, farmers switched crops and sold cattle to prepare for the dry spell.[9]

Such preparations can make a huge difference in how a particular area weathers the event. When a drought was forecast in December 1991, Brazilian officials built a dam

The photo above shows Its Beach in California in mid-December 1997 during typical winter conditions. Below, wind-whipped waves caused by El Niño weather pound the beach on February 8, 1998. Damage to the California coastline was less than during previous storms, however, because residents took steps to limit erosion after being warned about El Niño.

to conserve water, limited the amount of water residents used, and convinced farmers to plant fast-growing crops they could harvest before conditions got too dry. Though the drought struck in 1992, the nation's farmers produced 530,000 metric tons of grain, only 18 percent below their usual harvest. A 1987 drought that struck without warning had far worse consequences. That year, farmers harvested only 100,000 metric tons of grain, 15 percent of their normal output.[10]

A recently released report estimates that accurate El Niño forecasts could potentially save U.S. farmers between $240 million and $266 million.[11] The report, conducted by NOAA, concluded that farmers could plant crops best suited to the weather predicted and adjust their growing season based on the forecast.

Peru has used wind and water reports from the Pacific since the mid-1980s to help farmers plan which crops to plant and when to plant them. When an El Niño is predicted—with expected dry conditions in the northern farming regions of the country—farmers plant fewer crops that require wet conditions, like rice, and more crops, like cotton, that tolerate dry conditions.

According to Jim Luyten, a research director at Woods Hole Oceanographic Institution, farmers saved "tens to hundreds of millions of dollars" by rearranging their crops and planting and harvesting schedule to accommodate El Niño.[12]

"El Niño research," said Michael Glantz, scientist at the National Center for Atmospheric Research, "is one of the bright spots on the scientific horizon, as we enter the twenty-first century."[13]

American farmers could save as much as $266 million by planning their crops and planting schedules based on accurate El Niño forecasts, according to a NOAA report.

PART OF NATURE'S RHYTHM

As with every El Niño this century, the 1997–1998 version provided researchers with much valuable information on the phenomenon. Scientists hope to expand their knowledge even further by adding to the existing monitoring system. One reason computer models fail in their predictions is the lack of enough data on ocean temperature and other measurements of the air and water, according to Bob Weller, a senior scientist at Woods Hole Oceanographic Institution.[14]

Observation sites around the globe may someday provide those vital measurements and help forecasters predict more precisely when El Niño will appear and how it will affect specific regions. An international program known as the Global Ocean Observing System has been proposed to gather more complete information on the oceans. It would operate much like the system that collects weather data around the world. Such a system is expensive, but scientists believe that long-term, detailed observations of the earth's oceans are needed to crack the El Niño mystery. "We'll never understand the ocean that well," said researcher Luyten, "unless we can sample it as extensively as we can the atmosphere."[15]

Like the seasons, El Niño seems to be part of the rhythm of the earth's climate. It may seem that the event's cycle is irregular judging from its appearances during the last two centuries—every two to seven years. But it's possible that El Niño operates on a cycle based on hundreds or thousands of years.

If scientists can figure out El Niño's timetable, they can teach people how to adapt to the drastic weather changes that accompany it. Scientist Richard Barber of Duke University's Marine Laboratory says observations of past El Niños have taught us that "El Niño is not a disaster, anomaly, or cruel twist of fate; it is how Earth works."[16]

By studying the past, scientists hope to catch a glimpse of the future. Perhaps they can learn what causes El Niño. Perhaps they will detect a pattern that will make it easier to predict the weather cycles. So far, El Niño has proved to be a good teacher. Scientists hope that future events may well provide some of the remaining pieces of the puzzle.

source notes

EL NIÑO: FACT AND FANTASY

1. Barry, Dave, "Clueless," *Miami Herald*, Dec. 28, 1997.
2. "Not Enough Jobs to Go Around for Few Republicans Running: Topsy Turvy MA Elections," *Boston Globe*, March 8, 1998.
3. Glantz, Michael H., *Currents of Change: El Niño's Impact on Climate and Society* (New York: Cambridge University Press, 1997), 66.

CHAPTER ONE: WHAT IS EL NIÑO?

1. Interview with Jill Cournoyer, March 19, 1999.
2. Ibid.
3. Blom, Eric, "A Storm Beyond Compare," *Portland Press Herald*, Web site www.portland.com, January 1998.
4. "Insurers Preparing for Ice Storm Claims," *Boston Globe*, Jan. 14, 1998.
5. Barsugli, Joseph J., et al., "Effect of the 1997–98 El Niño on Individual Large-Scale Weather Events" (Boulder, CO: NOAA-CIRES Climate Diagnostics Center, at Web site http://www.cdc.noaa.gov/~jsw /Manuscripts/jjb_etal.html).
6. Interview with David Adamec, National Atmospheric and Space Administration (NASA).
7. Glantz, *Currents of Change*, 3.
8. Ibid., 111.
9. Ibid., 40.
10. Ibid., 123.

11. Suplee, Curt, "El Niño, La Niña: Nature's Vicious Cycle," *National Geographic*, vol. 195, no. 3, March 1999, 73.

12. "Adjusting to a Sea Change in the Weather," *Boston Globe*, Feb. 17, 1998.

CHAPTER TWO: REACHING ACROSS THE GLOBE

1. Williams, Jack, "What Is El Niño?" USA Today Information Network, June 12, 1998, at Web site http://www.usatoday.com/weather/nino.

2. Stevens, William K., "Remember El Niño? His Sister Has Shown Up and She's Angry," *The New York Times*, Jan. 27, 1999.

3. Nash, J. Madeleine, "Fire and Rain: El Niño's Fury Can Still Be Seen in the Drought and Raging Blazes of Borneo," *Time Science*, vol. 151, no. 15, April 20, 1998.

4. Brenner, Jim, "Southern Oscillation Anomalies and Their Relation to Florida Wildfires," Midwestern Climate Center, Champaign, Ill., at Web site http://flame.doacs.state.fl.us/Env/enso.html, March 2, 1998.

5. Ibid.

6. Siegel, Barry, "The Chaos of Nature—A Catastrophe Named 'El Niño,'" *Philadelphia Inquirer*, Aug. 21, 1983, G01.

CHAPTER THREE: EL NIÑO THE CELEBRITY

1. Siegel, "The Chaos of Nature."

2. Ibid.

3. Glantz, *Currents of Change*, 60.

4. Siegel, "The Chaos of Nature."

5. Ibid.

6. Ibid.

7. Ibid.

8. Ibid.

9. Amaral, Kimberly, "1982–1983 El Niño: The Worst There Ever Was," at Web site http://www.umassd.edu/public/people/kamaral/thesis/1982-1983El Nino.html.

10. Siegel, "The Chaos of Nature."

11. "Effects of the 1982–1983 ENSO," *The New York Times*, August 2, 1983.

12. Siegel, "The Chaos of Nature."

13. "Effects of the 1982–1983 ENSO," *The New York Times*, August 2, 1983.

14. Glantz, *Currents of Change*, 20–21.

15. "Effects of the 1982–1983 ENSO," *The New York Times*, August 2, 1983.

16. Ibid.

17. Amaral, Kimberly, "1982–1983 El Niño: The Worst There Ever Was."

18. "As the World Turns—A Surprise Gift from El Niño: Days Were One-fifth of a Millisecond Longer," *Philadelphia Inquirer*, Dec. 18, 1983.

19. Glantz, *Currents of Change*, 20–21.

20. Ibid., 169.

21. Ibid.

CHAPTER FOUR: TRACKING EL NIÑO

1. Siegel, "The Chaos of Nature."

2. Nash, "Fire and Rain: El Niño's Fury Can Still Be Seen in the Drought and Raging Blazes of Borneo."

3. Glantz, *Currents of Change*, 161.
4. Twain, Mark (Samuel Clemens), "New England Weather;" speech to the New England Society, Dec. 22, 1876, in *Bartlett's Familiar Quotations*, (Boston: Little, Brown, and Co., 1968), 759.
5. Glantz, *Currents of Change*, 89.
6. ———, "Forecasting El Niño: Science's Gift to the 21st Century," Environmental and Societal Impacts Group, National Center for Atmospheric Research, at Web site http://www.dir.ucar.edu/esig/elnino/glantz1.html.
7. ———, *Currents of Change*, 130.
8. Ibid., 79.

CHAPTER FIVE: EL NIÑO 1997-1998

1. Associated Press, "Buoys Aid in El Niño Prediction," in *San Diego Union-Tribune*, Feb. 13, 1999, B–10.
2. Interview with David Adamec.
3. Environmental News Network staff, "El Niño Observed from Start to Finish," Jan. 6, 1999, at Web site http://www.enn.com/specialreports/elnino.
4. Suplee, "El Niño, La Niña: Nature's Vicious Cycle," 88.
5. Koop, David, "Weather-beaten Peru Feels Brunt of El Niño's Rage," Associated Press, in *San Diego Union-Tribune*, March 30, 1998, A–1.
6. "Sand Slowly Disappearing from LA-Area Shoreline," *Los Angeles Times*, July 4, 1998, at Web site http:// www.latimes.com.
7. "Big Island Endures El Niño Drought," *Los Angeles Times*, Feb. 22, 1998.

8. "Flooding in Kenya Kills at Least 86," *Los Angeles Times*, Jan. 18, 1998.

9. "38 Dead in Worst Florida Twister Attack," *Los Angeles Times*, Feb. 24, 1998 (later updated to 40).

10. "Drought Raises Starvation Fears," *Boston Globe*, Sept. 10, 1997.

11. "Flooding in Somalia Destroys Crops, Homes," *Boston Globe*, Nov. 12, 1997.

12. "Group Says 1997 Was 'The Year the World Caught Fire,' " *Boston Globe*, Dec. 17, 1997.

13. "Ill Wind from Guam," *Boston Globe*, Dec. 21, 1997.

14. Nash, "Fire and Rain: El Niño's Fury Can Still Be Seen in the Drought and Raging Blazes of Borneo."

15. Associated Press, "FEMA Says Previous Winters Were More Expensive: El Niño Not So Costly," *ABC News*, April 6, 1998, on Web site ABCNews.com.

16. "Ski Slopes and Woods Show Some Wear and Tear," *Boston Globe*, Jan. 13, 1998.

17. Crossen, Cynthia, "El Niño Gets El Blame-o for Almost Everything," *The Wall Street Journal* in the *San Diego Union-Tribune*, April 27, 1998, E–1.

18. Suplee, "El Niño, La Niña: Nature's Vicious Cycle," 73.

19. Abramovitz, Janet N., and Seth Dunn, "The Toll of Natural and Not-so-natural Disaster," Worldwatch Institute in *San Diego Union-Tribune*, Dec. 3, 1998, B–11.

20. Associated Press, "Buoys Aid in El Niño Prediction."

21. Borenstein, Seth, "Drought Could Be in Works—

Dust Bowl-size Dry Spell Expected Soon," Knight Ridder News Service, in *San Diego Union-Tribune*, Dec. 16, 1998, A–9.

22. Tangonan, Shannon, "La Niña Turmoil Likely to Continue Through June," *USA Today*, Jan. 26, 1999.

23. Ibid.

24. McNeil, Donald G. Jr., "Wild Weather Takes Its Toll in Once-mild South Africa," New York Times News Service, in *San Diego Union-Tribune*, Dec. 26, 1998, A–22.

25. "SE Asia Awaits Monsoon to Smother Smoggy Fires," *Miami Herald*, Sept. 27, 1997.

26. Maniam, Hari S., "Asia on Alert for Deadly Haze," Associated Press, in *Journal Tribune*, Sept. 24, 1997, 4.

27. "Loggers' Fires in Indonesia Decried as Health Disaster," *Boston Globe*, Sept. 26, 1997.

28. Abramovitz and Dunn, "The Toll of Natural and Not-so-natural Disaster."

29. Environmental News Network, "El Niño Offers Peek at Global Warming," Oct. 19, 1998, at Web site http://www.enn.com/specialreports/elnino.

30. Ibid.

CHAPTER SIX: FORECAST FOR THE FUTURE

1. Graham, David E., "Unusually Dry Winter Is Forecast for County," *San Diego Union-Tribune*, Oct. 23, 1998, B–1.

2. "Red Cross Sets New Goal, Focus," *Boston Globe*, Jan. 8, 1998.

3. Gorov, Lynda, "Californians Say Live and Let Dry," *Boston Globe*, Feb. 11, 1998, A4.

4. "Weather Worries Lead Peru to Borrow," *Miami Herald*, Sept. 24, 1997.

5. "Strange Brew: Bringer of Bounty and Famine, El Niño Keeps Experts Guessing," Cable News Network at Web site http://cnn.com/specials/el.nino.

6. "US Adopts New Limits on Pacific Ground Fish," *Los Angeles Times*, Dec. 29, 1997.

7. "Mexicans Got Warnings; No Hurricane Deaths Cited," *Boston Globe*, Nov. 11, 1997.

8. Suplee, "El Niño, La Niña: Nature's Vicious Cycle," 85.

9. "Strange Brew: Bringer of Bounty and Famine, El Niño Keeps Experts Guessing."

10. Glantz, *Currents of Change*, 80–81.

11. Environmental News Network, "Study Tallies Worth of El Niño Forecasting," March 23, 1998, at Web site http://www.enn.com/specialreports/elnino.

12. Nadis, Steve, "Unraveling El Niño: A Conversation with Jim Luyten," *Woods Hole Currents*, vol. 7, no. 4, Fall 1998 (Woods Hole, MA: Woods Hole Oceanographic Institution, 1998), 5.

13. Glantz, *Currents of Change*, 137.

14. Nadis, Steve, "Chasing the GOOS (Global Ocean Observing System)," *Woods Hole Currents*, vol. 7, no. 4, Fall 1998 (Woods Hole, MA: Woods Hole Oceanographic Institution, 1998), 7.

15. Nadis, "Unraveling El Niño: A Conversation with Jim Luyten."

16. Glantz, *Currents of Change*, 166.

glossary

anchovy
Small, coldwater fish that resemble herring and that abound off the coast of Peru; used for food and fish meal.

anomaly
Unusual happening or change from the normal condition.

atmospheric scientist
An expert who studies the layers of air encircling the earth.

barometer
A device used to measure atmospheric pressure.

biologist
A scientist who studies living things.

climate
The weather conditions that are typical in a certain region.

computer model (also mathematical model)
A computer setup that duplicates the expected behavior of a particular event, for example, that of El Niño; used to predict the onset of El Niño and the effect it will have on the earth's weather. Scientists use mathematical formulas to create such models and to direct their operation.

conquistadors
Explorers from Spain or Portugal who conquered the

tribes living in Mexico, South America, and Central America during the 1500s and 1600s.

drought
An exceptionally long period with no rain; usually kills crops or seriously reduces the harvest in affected areas.

El Niño
Spanish for little boy or Christ Child; a warm pool of water that forms along the equator off the western coast of South America every two to seven years. Also referred to as a warm event.

encephalitis
An often-fatal illness, spread by mosquitoes, that causes the brain to swell.

ENSO
El Niño-Southern Oscillation. The event that occurs when waters in the eastern Pacific Ocean along the equator heat up and easterly winds along the equator shift direction.

famine
Massive starvation throughout a region.

fish meal
Ground fish used to feed cattle and poultry; also used as fertilizer.

fossil fuel
Fuel created from the remains of plants and animals pressed

under the soil for millions of years; oil and coal are examples of fossil fuels.

global warming
A gradual increase in the earth's temperature, believed by many scientists to be caused when pollution from burning fossil fuels traps heat in the atmosphere.

guano birds
Coastal birds, found in Peru and other areas, whose dung is used to make fertilizer.

hurricane
A severe storm created in the tropic regions of the Atlantic Ocean or the Caribbean Sea; hurricanes produce heavy rains and winds exceeding 75 miles (120 km) per hour.

hydrographer
A scientist who studies and maps bodies of water.

jet stream
A narrow corridor 25,000 to 35,000 feet (7,620 to 10,700 km) above Earth, along which high-speed winds blow from west to east at speeds up to 200 miles (320 km) per hour.

Kelvin Wave
A gigantic underwater wave that can be created when winds change direction over the equatorial Pacific Ocean.

La Niña
Spanish for little girl; seen as the opposite of El Niño. Cold

pool of water that flows from west to east across the Pacific and along the equator; it can bring weather conditions that are opposite to those produced during El Niño; also referred to as a cold event.

Lyme disease
An illness caused by a virus spread by deer ticks; causes joint pain and flulike symptoms.

malaria
An often fatal disease carried by mosquitoes that causes chills, fever, and sweating in victims.

meteorologist
An expert on the earth's atmosphere and the weather.

monsoon
Steady winds, particularly in India and Asia, that are responsible for alternating seasons of rain and dryness.

National Oceanic and Atmospheric Administration
(NOAA) The U.S. government agency that oversees activities and research relating to the oceans and the atmosphere; a branch of the Commerce Department.

National Weather Service (NWS)
A division of NOAA that oversees weather research, monitors weather events, and issues weather reports.

oceanographer
A scientist who specializes in the study of the ocean.

radar
High-frequency radio waves used to locate and study far-away objects; the word is short for RAdio Distance And Ranging.

Southern Oscillation
A pattern of atmospheric pressure over the east and west tropical Pacific Ocean that swings back and forth between high and low.

Tropical Atmosphere Ocean array (TAO)
A system of 70 buoys strung together across the Pacific Ocean that record sea conditions and transmit the information via satellite to weather stations around the world.

thermocline
The layer of water that separates the cold depths of the ocean from the warm top layer of water.

thermometer
A device that measures temperature.

Tropical Ocean Global Atmosphere (TOGA)
A program set up by the United Nations in 1985 to monitor weather conditions, using satellites, moored and drifting buoys, and tide gauges.

TOPEX/Poseidon satellite
A satellite launched in 1992 that uses radar to map the ocean's surface height or level; TOPEX stands for TOPography EXperiment.

tornado
A funnel-shaped column of air that whirls at speeds of up to 300 miles (480 km) an hour, causing destruction along its path.

trade winds
The major winds that blow along the equator in a regular pattern to the southwest in the Northern Hemisphere and to the northwest in the Southern Hemisphere.

typhoon
A hurricane in the west Pacific Ocean or the China Sea.

upwelling
Upward movement of cold, nutrient-rich ocean water.

Walker Circulation
The circular motion of air above the tropical Pacific Ocean from east to west at the lower levels and from west to east in the upper atmosphere; named for Sir Gilbert Walker.

weather
The atmospheric conditions of a specific place on a certain day.

weather satellite
A space device that circles the globe and monitors and transmits signals that report weather conditions on earth.

for further information

BOOKS FOR STUDENTS

Can It Really Rain Frogs: The World's Strangest Weather Events (Spencer Christian's World of Wonders) by Spencer Christian and Antonia Felix, John Wiley & Sons, 1997. A book for all ages on amazing weather phenomena.

Changes in the Wind: Earth's Shifting Climate by Margery Facklam, Marjorie F. Facklam, Paul Facklam, and Howard Facklam, Harcourt Brace, 1986.

El Niño: Stormy Weather for People and Wildlife by Caroline Arnold, Clarion Books, 1998. Clearly written account of El Niño, how it forms, scientists' efforts to track it, and its effect on the earth.

Hurricane (A Disaster! Book) by Christopher F. Lampton, Millbrook Press, 1991. Describes hurricanes, the destruction they cause, and how they are detected, measured, and predicted.

Hurricanes (The Weather Channel) by Susan Hood, Simon Spotlight, 1998. Uses dramatic color photos to describe hurricanes and explain basic weather information.

Hurricanes: Earth's Mightiest Storms by Patricia Lauber, Scholastic, 1996. Eyewitness accounts, scientific facts, and full-color photographs that explain the formation of hurricanes and their effect on those in their path.

The New Book of El Niño: With Computer Reconstructions and Artists' Impressions of the Latest Weather Patterns Around the World by Simon Beecroft, Copper Beech Books, 1999. Readers explore the reasons behind recent weather conditions that are causing problems worldwide.

Tornadoes (First Book) by Ann Armbruster and Elizabeth A. Taylor, Franklin Watts, 1993. Tornadoes are described in vivid detail in this clearly written book; full-color photos and simple experiments.

Weather (National Audubon Society First Field Guide) by Jonathan D. Kahl, Scholastic Trade, 1998. Full-color photo spreads explain the weather in part 1 of this guide. Part 2 is a field guide with detailed descriptions of every conceivable weather condition. Part 3 lists references for further information and includes a waterproof spotter's guide.

Weather Watch: Forecasting the Weather (How's the Weather?) by Jonathan D. W. Kahl, Lerner Publications, 1996. This book combines a history of meteorology with a discussion of weather, including a section on forecasting; color photos, maps, and illustrations.

Weather: Mind-Boggling Experiments You Can Turn into Science Fair Projects (Spectacular Science Projects) by Janice Pratt Vancleave, John Wiley & Sons, 1995. Students learn about the weather through these easy-to-perform experiments.

ADULT BOOKS

Currents of Change: El Niño's Impact on Climate and Society by Michael H. Glantz, Cambridge University Press, 1996.

WEB SITES

http://abcnews.go.com/sections/science/dailynews/elnino_how.html ABC News site with news stories, graphs, charts, maps, and explanations of El Niño and La Niña.

http://airsea-www.jpl.nasa.gov/ENSO/welcom.html
NASA's Jet Propulsion Laboratory's page on El Niño and the monitoring system used to predict the event.

http://cnn.com/SPECIALS/el.nino Cable News Network's special Web site devoted to El Niño. With video, photos, news stories, maps.

http://topex-www.jpl.nasa.gov Topex/Poseidon home page.

http://www.elnino.noaa.gov NOAA's homepage on El Niño. Graphics, maps, and charts; information on forecasts, observations, and research; links to many other sites.

http://www.enn.com/specialreports/elnino Special site on El Niño operated by the Environmental News Network; graphics, links to other sites, news reports.

http://www.nationalgeographic.com/elnino Site on El Niño/La Niña operated by National Geographic with comments from readers on how El Niño affected them and article on ENSO. Also links to other sites.

http://www.ogp.noaa.gov/enso El Niño-Southern Oscillation home page, operated by NOAA's Office of Global Programs. Comprehensive site on El Niño and La Niña with graphics, maps, and charts; also links to other sites including reports in French, Portuguese, and Spanish.

http://www.pmel.noaa.gov/toga-tao/el-nino/nino-

home-low.html NOAA's site on El Niño—Detailed look at El Niño; lots of colorful graphics, links to more information. A project of Department of Commerce / National Oceanic and Atmospheric Administration/Pacific Marine Environmental Laboratory / Tropical Atmosphere Ocean Project.

http://www.pmel.noaa.gov/toga-tao/home.html NOAA's TAO Project Home Page.

http://www.usatoday.com/weather/nino/wnino3. htm This is *USA Today*'s special site on El Niño with many facts on the event and weather as well as the history of El Niño.

index